BRITAIN'S HERITAGE

# Steam Lorries

*Anthony Coulls*



AMBERLEY

# Acknowledgments

I would like to thank Peter Coulls and Hedd Jones for proof reading, comments and suggestions. Thank you too to Chris Mayall, Derek Billings and Ian Cooper for specific photographs of Aveling, Wallis, Tasker and Sentinel lorries. Thanks also to Nick Wright for use of his photograph. Other photographs were taken by myself or Peter Coulls, or are from our collections or the collection of Hedd Jones.

This book is dedicated to the memory of the late Mike Jones, stalwart of steam lorry preservation and research, whose photograph of Sentinel 8122 at work in North Wales in the summer or 1977 is on the front cover; also to the memory of the late Tony Cooper, a true Salopian with a great enthusiasm for the products of the Sentinel works where his grandfather had been chief superintendent of the metalwork shop.

***Cover***: Sentinel Tarsprayer 8122 in the ownership of Lloyd-Jones brothers at work in North Wales during the summer of 1977.

First published 2020

Amberley Publishing
The Hill, Stroud
Gloucestershire, GL5 4EP

www.amberley-books.com



ISBN 978 1 4456 9850 2 (paperback)
ISBN 978 1 4456 9851 9 (ebook)



British Library Cataloguing in Publication Data.
A catalogue record for this book is available from the British Library.

Typesetting by Aura Technology and Software Services, India. Printed in the UK.

# Contents

# 1

# The Beginnings of Steam Road Transport

In the twenty-first century, steam power is seen very much as something from the past. For all its long history, it remains a contemporary form of energy and steam is a key part in the production of hydrogen for hydrogen fuel cells, now at the cutting edge of technology. Steam itself, water vapour under pressure, is very clean – it's the method of production, in terms of steam raising with a boiler, where there is the possibility of smoke or flue gases. At the end of the nineteenth century, there were not so many concerns about pollution, climate change or greenhouse gases, and the high technology development of traditional steam power was itself driving the industrial United Kingdom forward. Steam power led the transport revolution in England throughout the century, but was crippled on the country's road network by restrictive legislation. As the century turned, the laws were altered in such a way that allowed the development of the English Steam Lorry or Wagon into a viable transport proposition. Speed and ease of use were key considerations and in the heritage world of the present day, the contrast on the steam rally field between traction engines and steam rollers on one hand and the steam lorries is marked indeed. For the best part of four decades, the steam lorry was a major player on the transport scene, being innovated and developed into a highly technical machine designed to beat competition from the petrol and diesel lorry as well as fighting the onslaught of further laws in the twentieth century. Made by a variety of builders, including the famous Sentinel company of Shrewsbury, who built waggons with a double g, steam lorries came in many shapes and sizes and this book will look at these in due course. First, though, let us take a look back in history to set the scene for the development of the self-propelled steam road vehicle.

Steam on the common roads of Britain had an uneasy start. From the late eighteenth century, steam traction powered the industrial and transport development of the British Isles. Whilst railways grew and adopted steam locomotion, the poor state of the road network – such as it was – along with punitive legislation, hampered the widespread use of steam-propelled vehicles for serious commercial use other than heavy haulage, road making or agricultural purposes. The principle of steam driving a wheeled vehicle was well known and proven. Nicolas Cugnot in France had built his Fardier steam tractor in 1769 and, incredibly, this machine still survives as an exhibit in the Musée des Arts et Métiers in Paris and has been a museum piece for over 200 years! By the 1780s, the concept of mobile steam locomotion was in Britain and being trialled by William Murdoch, albeit with test model carriages, but these models paved the way for further experimentation and eventual implementation. In the early nineteenth century, steam road carriages were developed by Richard Trevithick, pioneering in his use of high-pressure steam – initially with the mobile *Puffing Devil* but moving on to ones to carry people in some degree of comfort. Sir Goldsworthy Gurney and William Hancock progressed with commercial steam carriages in the 1830s for the

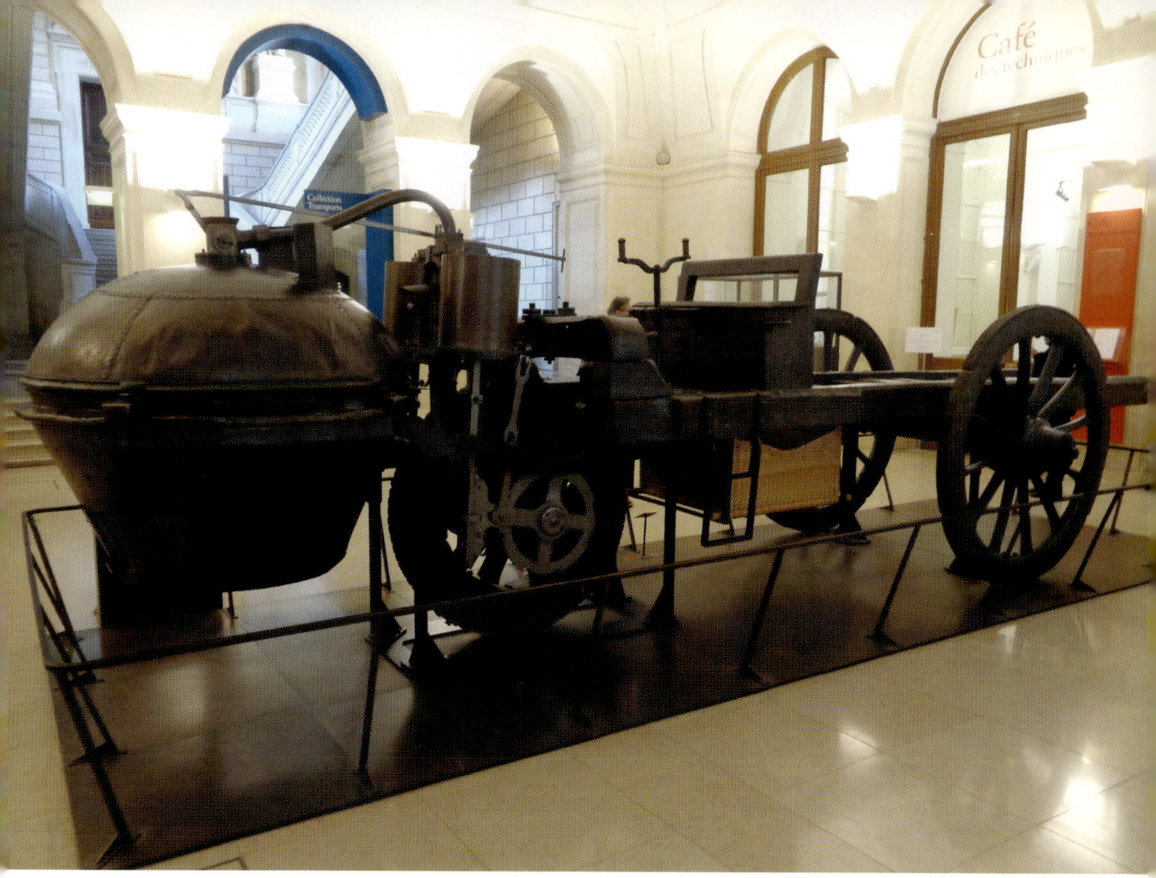


Nicolas Cugnot's steam cart was made in 1769 and was surprisingly effective despite managing a collision in its short working life. Since 1800, it has been an exhibit in the Musée des Arts et Métiers in Paris. Well worth seeking out, as I did in November 2019, it is the oldest self-propelled vehicle in the world.

conveyance of passengers, and the engine of a Gurney steam drag (as it was known) can be seen in Glasgow Museum of Transport. The first half of the nineteenth century saw great progress in steam vehicle design on both road and rail, and by the 1850s it was also used to produce locomotives on a commercial basis. Progress was dampened by legal restrictions – mainly the punitive Red Flag Act of 1865 – which limited or prohibited the use of steam-powered vehicles on roads, and it was not until the 1860s that the idea of actually carrying the commercial load on the vehicle itself began to be investigated.

Following a relaxation in the legislation covering the use of steam-powered vehicles on common roads, which included lifting speed limits, applying weight restrictions and, importantly, removing the need for a man to walk in front carrying a red flag, manufacturers started to investigate the possibility of using steam power for a self-contained goods vehicle. Prior to this point, goods were carried in a trailer towed behind a traction engine, or more frequently a horse. The scene was set for the arrival of the steam lorry, or lurry, as some of the reports or builders called them at the time.

The laws of the land that severely restricted the unladen weight of wagons challenged designers and manufacturers to be creative in their meeting a solution to the practical transport need. Steam wagon production began to flourish in the UK in the last decade of the nineteenth century after the repealing of the Red Flag Act in 1896. Manufacturers such

**The first Thornycroft van weighed 1½ tons with an engine made by Simpson Strickland & Company. Dating from 1896, it is currently housed at the Historic Commercial Vehicle Museum in Leyland, Lancashire.**

as the Lancashire Steam Motor Company (who would become Leyland), Coulthard, Mann, Straker and Thornycroft were among the companies that began producing wagons at this time. The first recorded steam-powered commercial vehicle of this period was actually an oil-fired van. Thornycroft were the builders, from Chiswick in London, and the van from 1896 remains, now on display at the British Commercial Vehicle Museum in Leyland, Lancashire. Thornycroft had been used to making steam plants for small boats, so to transplant such a plant onto four wheels and an underframe made perfect sense. It laid the way for light steam development in a separate way to the traction engine and road locomotive. These latter pulled trailers – a new development which allowed the goods to be carried on the powered vehicle itself, setting the standard for future development of the steam lorry, and indeed the petrol and diesel lorries after that. A series of Heavy Motor Car Trials under the auspices of the Liverpool Self-propelled Traffic Association were held in 1898, 1899 and 1901 gave the new light steam lorry a chance to be shown to the world at large. Fifty invitations were sent out to attend, but in the event, only three makers brought their products to meet the challenge. Competing for a gold medal, the vehicle had to haul a weight of 2 tons around the course – and only steam vehicles were able to do this initially, as the petrol engine was not developed enough.

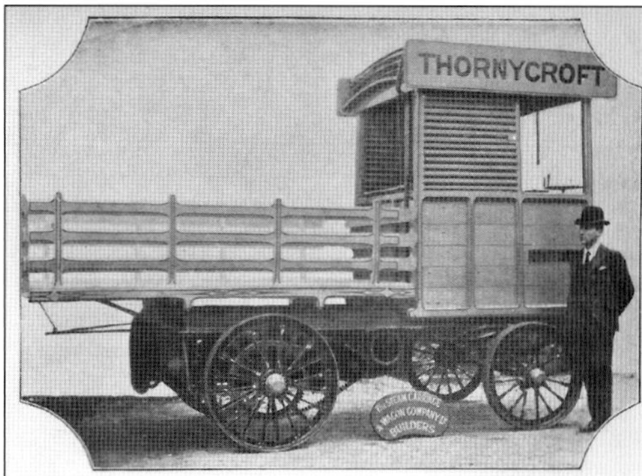


The "THORNYCROFT" Four-wheeled Lorry.

The "THORNYCROFT" Six-wheeled Lorry.

**From the 1899 trials, two images of Thornycroft's entrants. Notice how light they are in construction compared to the later steam lorries.**

The first trial was entered by the Lancashire Steam Motor Company, Thornycroft and Lifu. Lifu was an abbreviation for Liquid Fuel, for their lorry, which burned paraffin; unusually for a manufacture of steam road vehicles, Lifu came from the Isle of Wight. In this trial, victory was claimed by the Leyland lorry, which won £100. The next trial still saw a little competition from the petrol engine, and of the thirteen vehicles presented for competition, eight were steam-driven. Eight different makers were represented; the gold medals went to Leyland, Thornycroft and Coulthard. Throughout the period that the steam lorry was a commercially viable enterprise, coal remained the primary source of energy – oil burning never really took off despite the initial success of the Lifu company. The trials were assiduously reported in the press of the time; full details of the contenders, conditions and results were set down in *The Engineer* journal, providing a fascinating insight into the innovative features of the lorries. For example, it was noted that the Coulthard lorry's boiler could have its inside removed without disturbing any of the steam fittings, and like the Lifu, it burned liquid fuel. It's interesting to note that the journal felt that more reliable entries would be received if a larger entrance fee had been set for the trials, with a portion being returned on the appearance of the vehicle for the trials!

In 1901, several makers competed in the Aldershot trials for the War Department, with Thornycroft's gear-driven undertype coming out as the winner ahead of Foden of Sandbach's early chain-driven overtype. The details of these layouts of lorry are explained shortly in the book. Both manufacturers built on this early success, with Foden patenting the essential features of the overtype wagon and deterring other manufacturers from attempting such a design. Despite this, Naylers of Hereford built overtype steam lorries from 1903 to 1909 whose compound design and general layout would have resulted in Foden's reaching for their copyright lawyers in the twenty-first century. Even the American manufacturer of Buffalo Springfield made steam lorries with a peculiarly English look about them, with more than a hint of Cheshire influence – most of the American steam practice had a very distinctively transatlantic look. In 1906, Wallis & Steevens produced an overtype wagon that Foden viewed as an infringement of their patent. The matter led to a patent infringement case. In 1908 the matter was decided in Wallis & Steevens' favour, and upheld on appeal.

***Above***: The Lifu lorry of 1899, seen in a paper discussing the Liverpool trials of light commercial vehicles. The layout of the undertype steam lorry is already becoming clear.

***Left***: Apart from the steam engine style chimney, the Clarkson lorry has much more visually in common with the petrol engine rivals it was built to compete against. Unfortunately, the company did not endure in the steam market.

This naturally led to a great expansion of overtype wagon production, with prominent traction engine companies drawing on their experience building steam tractors to produce wagons, with varying success.

In 1901 the Yorkshire Patent Steam Wagon company began producing undertype wagons in Leeds with their distinctive pattern of double-ended boiler. Whilst this looked unconventional, the basic design carried on for thirty-six years. The novel double-ended transverse-mounted boiler was used to avoid problems of tilting when climbing hills. Internally it resembled a locomotive or Fairlie boiler with a central firebox and multiple fire-tubes to each end. In the Yorkshire though, a second bank of fire-tubes above returned to a central smokebox and a

***Above***: By 1900, the Thornycroft lorry looked like *Dorothy*, a very unusual survivor from the beginning of the century. Taken into preservation in the late 1950s, the lorry is seen at a former Bedford steam rally, held at Old Warden. ***Right***: Wallis & Steevens of Basingstoke are well known for their steam rollers, traction engines and steam tractors. Like many other makers, they tried their hand at making steam lorries, and this 1912 example in Milestones Museum, Basingstoke, is a unique survivor in the United Kingdom.

The form of the Foden overtype was quickly established in the 1900s, and the market spread wide. In 2008, this 1907 example was newly restored and shown at an event at Quex Park, Kent. It had spent its working life in Chile until it returned to the UK for restoration.

single chimney. In this period, many manufacturers made attempts to build steam wagons, often with only moderate success, and the book will turn to them in due course.

The 1904 Heavy Motor Car Order resulted in Foden amending their design of 5-ton lorry and producing a lighter 3- to 4-ton lorry. The Order laid down various requirements on speed, load and tyres and the 3-ton lorry could take full advantage of this. Fitted with rubber tyres and being able to run at 12 miles an hour due to a faster drive arrangement, the load being carried remained economically practicable.

The Yorkshire Patent Steam Wagon Company of Leeds made very distinctive vehicles with their unique transverse boilers. Very successful machines, the London haulage contractors Beck & Pollitzer ran a number of Yorkshire lorries in their large fleet.

The 3-ton Foden is seen to advantage in this very finely restored 1912 lorry, owned by Ian Calder and seen when very newly outshopped in November 2016 at Statfold Barn in Staffordshire.

# 2

# Undertype, Overtype? Waggoning Free...

Throughout the age of commercial steam in agriculture and in heavy road haulage, most traction engine and road locomotive construction took place in England. Perhaps peculiarly, no engines were produced in the traditional industrial heartlands of London, Birmingham or Manchester – but this was not the case with the building of steam lorries. A good number of the existing traction engine builders entered the market, notable exceptions being Marshall, McLaren, Ruston and Greens, whilst other firms such as Atkinson, Thornycroft, Sentinel, Yorkshire and Leyland built nothing but lorries. Many smaller firms entered the field in the early years of steam lorry development, with greater or lesser degrees of success; some vanished quickly, others diversified and flourished, but new locations of manufacture were seen, as noted above.

In the initial period of development, two main directions were followed by way of layout and technical innovations. These were the overtype, in which the engine unit was mounted on top of a steam boiler at the front of the lorry, and the undertype, in which the engine was located under the floor of the lorry, leaving more space for the payload. As time wore on, some manufacturers experimented with both types, but the two eventual market leaders stayed with the designs that were popular and worked for them – Fodens of Sandbach sticking to overtypes in the main and Sentinel developing and enhancing the undertype for over three decades. Chain drive to the back axle gave way to shaft drive in the later Sentinels, but the initial basic layout remained. Allchin, Aveling & Porter, Burrell, Garrett, Clayton & Shuttleworth, Robey and Wallis & Steevens were amongst those who made overtype steam lorries in some number, though there are few survivors of most of these makers' vehicles surviving in the world, not just the United Kingdom.

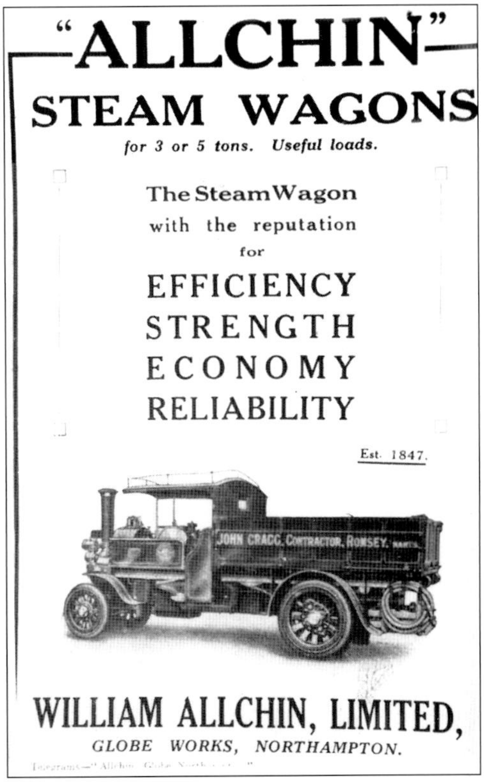


Advertising was all key when competition was fierce. Taking advantage of the 1904 Heavy Motor Car Act, Allchin tell their potential customers that both 3- and 5-ton models were available.

***Left***: Whilst Aveling & Porter made steam lorries in some quantity for the home market, the sole survivor was retrieved from Australia for restoration. Now resident in Lancashire, it was initially restored by the late Tom Varley, who was renowned for seeking out rare steam engines for preservation.
***Below***: Clayton & Shuttleworth made this overtype in 1920, typical of many of its make, but by 2019 an unusual survivor. It was photographed at the annual Great North Steam Fair held at Beamish Museum.

***Above***: The Garrett overtype lorry was effective and efficient, though survivors are few. Southern Transport, as with many other companies, ran a fleet of these, typical of the early twentieth-century contractors before serious competition from internal combustion vehicles.
***Below***: Manufacturers were rightly proud of their products. Robey of Lincoln took this photograph of one of their lorries outside the works when new and before delivery to the customer.

There were departures from these basic layouts, including attempts at front-wheel drive, locating the engine in different parts of the lorry or, like Fowler, using a 'V' layout engine behind the driver. Perhaps the most visually different set-up was that of the Yorkshire steam lorry. This has already been described in its style, but as with all the longer lasting players in the field, it was developed over time so that the last built lorries still had the same basic arrangement of features but were far more suited to the roads and traffic prevailing at the time.

The standard layout of steam lorries with conventional locomotive-type boilers was a variation on the traction engine concept with a compound or twin-cylinder steam engine mounted on top of the boiler itself – hence the name overtype. Drive from the crankshaft was through gears and chains and steering by worm, gear and chain, as with traditional traction engines, initially. Later lorries were developed with direct gear steering which was more positive, and refinements such as superheating added, depending on the maker. A chassis extending from the engine unit formed the load-carrying basis of the lorry. The actual operation of the steam engine is better explained in the companion book in this series, *Traction Engines*. The steam lorry was perhaps the most developed concept of the road steam vehicle apart from the steam car – and certainly, springs, good steering and an effective, economic engine and boiler unit were the elements to that success. It is fair to say that by the time the steam lorry was becoming widespread, no single-cylinder units were ever produced, unlike the steam roller and traction engine, where that configuration was still in use on some of the last steam rollers made in the UK in the early 1950s.

**The 5-ton Foden overtype could carry prodigious loads, albeit at the expense of speed. There's a fair amount of flour on the back of this lorry, photographed in Eynsham; the pride of the driver and his mate is evident.**

Once a proven power unit and chassis was created, the steam lorry could be adapted to almost any use the customer wished it to be put to. Box vans and flat beds were versatile for general carriers. The development of technology allowed tipping bodies to be created, both side tipping and rearwards, initially hand-operated but latterly by steam. Tanker bodies were made for movement of liquids – tar, oil and beer, just to name three. House removals by steam were possible, and the bulk movement of bales of material, cotton, sugar and flour were all regular loads for steam. Single lorries could be owned by an operator or an entire fleet, like those operated by household names such as Tarmac and Tate & Lyle. A steam lorry could pull a trailer – and often did – and the development of an articulated unit wasn't far behind, though there weren't many in use compared to their diesel counterparts. The hardy steam lorry could also be used to move stone in large quantities and Aberdeen Harbour Authority ran a six-wheeled wagon for sea defences and dock work. Some more specialist uses of steam lorries included gully emptying, tar sprayers and even livestock transport – there is a Sentinel works photograph of a pig van!

More elaborate and unusual uses of steam lorries included the steam bus – the most famous being one built on a Foden chassis that was used to transport the Foden Works Band. Known as *Puffing Billy*, it sadly did not survive into the heritage era, but a preserved Foden steam lorry was rebuilt as a bus to the same pattern in the 1960s and is now operating in Sussex for its enthusiast owners. In recent years a couple of Sentinel steam buses were operating out of Whitby and Morecambe; these no longer run, though the Whitby bus has been overhauled and is now used by Crosville of Weston-super-Mare for hire on special occasions. A further couple of steam buses include a Sentinel S-type and a Foden overtype of 1923 which runs in Cheshire and Shropshire – including taking the author's sister and

**Not recommended for everyday use, but the versatility of the steam lorry is seen in the immense and awkward load carried by this Sentinel DG6 on a civil engineering contract in Dorset.**

***Above***: A tipping body allowed easy unloading of goods, no doubt a key factor when brick and plaster makers Cafferata of Newark ordered this Sentinel DG6. It's captured demonstrating its capabilities before delivery to Nottinghamshire.
***Below***: Very few articulated steam lorries survive. In June 2009, this Foden tractor was owned by Stuart Ritchie who was restoring the 'fifth wheel' on the rear of the lorry to allow it to pull articulated trailers. It was attending the Corbridge steam rally at the time and is a fine example of the splendid restorations of recent years.

Displaying their headlamps and glass windscreens to the fore, some of the Garrett undertype lorries belonging to Gloucestershire County Council make a fine sight when moving roadstone in the inter-war period.

brother-in-law from their wedding to the reception. The idea of a steam bus for everyday service remained something from the earlier period of steam road transport; the commercial fleet of Clarkson steam buses that were run by the London General Omnibus Company were ousted in 1909 by new Metropolitan regulations on weight. By this time, a rival company was operating buses with Daimler petrol engines that were far more successful than they had been a decade earlier.

At the Great North Steam Fair in April 2017, the 1923 Foden, fitted in 1965 with the replica bus body, came up from Sussex to attend the event. Named *Puffing Billy* like the original, it no longer carries the Foden works band, however.

Commercial steam on the road in the twenty-first century is an unusual sight. For a period in the summer of 2005, a passenger service was offered by *Martha*, a Sentinel DG4 fitted with a bus body earlier that year. Running north from Lancaster, I captured it passing the Midland Hotel, Morecambe (itself under restoration), on 22 June 2005.

Even more high profile on the street than a steam bus was the use of steam lorries by showmen who travelled their fairground rides around the country. The large rides were towed by showmen's road locomotives with the ability to generate electricity to power the rides, but the versatility and speed of the steam lorry was quickly recognised. Smaller attractions, side shows and stalls such as coconut shies and shooting galleries were easily moved by steam lorry and once on the fairground site – or tober as it was known – the lorry could use its steam engine disengaged from road wheel drive to run a dynamo in order to produce energy for lighting the fair. In the same way, the modern showmen have lorries at the back of their rides and attractions generating power for driving and lighting the fair. Ornately painted, lined out and lettered brightly, they were valued pieces of fairground equipment, though not as widespread as the motor lorry after the First World War. As far as is known, these vehicles were worked so hard that none survived into preservation, although one Foden in Cheshire is fitted with a dynamo and gives an impression of what these lorries would have been like.

We are so used to seeing traditionally dark red or green, blue or black steam road vehicles that to see an accurately researched livery on a preserved steam lorry can be an eye-opener. On the British steam rally scene, there are orange, yellow and white Sentinels,

Derelict after a hard life on the fairground, the Wallis & Steevens overtype of J. & H. Williams Chairoplanes awaits its fate in the 1930s.

Amusement caterer Fred Smith's Foden *Prince* demonstrates the capability of the steam lorry for the showman in moving side stalls and living vans. Larger rides were still the preserve of the road locomotive until diesel tractors came in.

in the liveries of Castle Firebrick, Blue Circle Cement and Hovis bread respectively. Many others provide a real contrast: purple for Cadbury's, and some other striking liveries – even the black Tarmac roadstone lorries are a sight to behold with their bold but simple branding. It's a mark of the interest that owners have in their lorries that more and more well-restored vehicles are being turned out with historically correct body styles and paint schemes. When one sees such a machine on the road or at an event at somewhere like Beamish in a historic street setting, it is easy to imagine the impact that the steam lorry had on the psyche at the time.

Undertype lorries initially followed a number of different layouts, mainly varying in the type and location of the boiler, as the engine remained under the floor of the body. There could be horizontal or vertical boilers, and a variety of boiler designs – but it was the Sentinel of Alley & MacLellan that set the standard for the most successful and innovative developments of steam on the road in the twentieth century.

In 1906, Alley & MacLellan of Glasgow built their first 'Sentinel' steam lorry, driven by an underslung engine with poppet valves, and an effective superheated vertical cross watertube boiler. It was a revolutionary design, and immediately took a large share of the market. Whilst it was developed highly over the next thirty years, the basic layout remained the same for Sentinels until steam production ceased in the 1950s. Only a couple of lorries remain in preservation from the Scottish period of production, testimony, however, to the incredible longevity of the steam vehicle. The company had a dalliance with overtype waggon building, but these were not particularly successful and only seventeen were made in total.

**Stuart Gray presents his S4 Sentinel lorry in the orange livery of its commercial owners, Castle Firebrick, and it turns heads wherever it goes. In September 2014, it was attending the sixtieth anniversary road run of the National Traction Engine Trust in Bedfordshire.**

The Sentinel S6 tipper of Peter Walker and Diane Carney has been a stalwart of the Midlands and Northern steam rally scene. In striking Tarmac black livery, Peter and Diane's daughter Jeanie was driving it in the arena at the 2019 Shrewsbury steam rally.

From the initial move of Sentinels to Shrewsbury in 1915, the first production lorries were the Sentinel Standard type, continuing from being made in Scotland. These were followed by the Super Sentinel in 1923, DG (Double Geared) in 1927 and S types, plus four, six- and eight-wheel configurations, moving from chain and gear to shaft drive.

An Ancestor of Sentinel Waggons.
Experimental Waggon running in 1902.

Alley & MacLellan's prototype Sentinel steam lorry of 1906 shows how the format was evolving for the companies' products in future years. Shown in promotional literature, note the spelling of 'waggon' with two 'g's!

***Left***: A Sentinel Standard wagon can still be seen every day in central London! Made in 1923 by apprentices at the Sentinel works, it is displayed at the Science Museum in South Kensington and is a 1:8 scale model!
***Below***: The Sentinel Standard lorry at the Grampian Transport Museum is one of the few remaining Alley & Maclellan machines made in Glasgow. Built in 1914, members of the National Traction Engine Trust were inspecting it in its 101st year during a visit.

DG and S-type production overlapped each other through the 1930s, with over 800 DGs and 400 S types being built. As well as lorries, some of these were also available as haulage or timber tractors, the latter with steam winches on their rears to add to the versatility. Many had hard working lives and were worn out in relatively few years compared to the steam roller or agricultural engine.

**Right**: April 2015's Great North Steam Fair gave an opportunity to compare a Standard Sentinel on the left from 1917 with a 1928 Super model on the right.
**Below**: The Hazell family have the unique ability to exhibit all three-wheel configurations of S-type Sentinel. From left to right: the S8, S6 and S4.

## Did you know?

Sentinel sold steam waggons as opposed to lorries or wagons, and the spelling of 'waggon' with two 'g's is peculiar to Sentinel alone. Every other builder sold wagons!

Sentinels also looked overseas and exported fifty to Czechoslovakia in the 1920s before arranging to produce waggons under licence as Skoda-Sentinels. This was a contrast to the early years of development in the UK, when continental builders attempted to enter the English market with the products of Serpollet, Scotte and De Dion Bouton at the beginning of the twentieth century with varying degrees of success. The export of lorries from England was widespread, many still surviving in Australia, with a few others in far-flung parts of the world such as Sri Lanka, where a Robey six-wheeled articulated lorry and a four-wheeled sister were in use until the 1980s. The latter lorry has recently been restored to steam by engineering students and its companion is a very rare survivor indeed, not just because of its make, but by dint of it being an articulated vehicle.

As the technology of the steam lorry progressed, there were developments to make them more user-friendly and efficient in operation. Some of these worked; others were experimental and exciting, but failed to find commercial success. This was particularly true in the years leading up to the First World War, where companies tried to make their mark in the market to a greater or lesser extent. Names who vanished from the steam lorry market without trace apart from adverts and photographs include Ellis and the Maidstone wagons,

**Steam lorries worldwide! Colombo docks in Sri Lanka was the working home to a Robey lorry and tractor. Both are now preserved in the country and subject to restorations to working order.**

Londonderry and Straker to name but three. Economics weighed heavily against them in a world dominated by Sentinel and Foden, who also established a service network to assist their customers for spare parts and repairs.

In its working environment, a Straker lorry demonstrates its capacity to pull a trailer and enhance its attractiveness to the customer.

Steam lorries were popular with breweries. In Kent, it was the obvious choice of the Dartford Brewery to use the undertype lorry produced by Jesse Ellis of Maidstone.

Railway companies were quick to use the steam lorry for wider distribution of goods from stations. The Londonderry lorries, made in Seaham, County Durham, found use particularly with the Belfast & County Down Railway in Northern Ireland.

## Did you know?

Beyer Peacock of Manchester were locomotive builders who diversified and made a very small number of lorries. Despite the limited production, they experimented with new ideas and were proud of the ability to change the engine unit on their lorries in 11 minutes from arrival in the repair yard, photographing the process so they could publicise the effectiveness of their operation. Fortunately for the historian, these images survive in the collection of the Science and Industry Museum in Manchester.

The great transport demands of the First World War led to several of the premier wagon manufacturers – at the time, Sentinel, Clayton & Shuttleworth, Foden and Garrett – having almost their entire production ordered directly for the war effort. This opened up the home market for many other manufacturers to fill the vacuum. A company that entered the market in this period was Atkinson, with their undertype wagon design launched in 1916. Atkinson of course would go on to remain in road vehicle production as Seddon Atkinson long after their last steam lorry hit the road. As such, they survived long enough to become part of Iveco, UK manufacture only ceasing in 2009.

The First World War was a major user of steam traction in conflict, and Foden lorries were sent out to France in large numbers. However, smoke by day and sparks by night made them ready targets for artillery and the casualty rate was high. A contemporary description of driving a steam lorry to the Front in a convoy taking supplies was published in the National Traction Engine Trust's magazine *Steaming* several years ago and is a harrowing read. The writer of the piece saw the lorry in front of him take a direct hit and the blast took the

Atkinson were a successful manufacturer who made the crossover into internal combustion lorries. There were no survivors in the UK until 1976, when Tom Varley imported this 1918 example and restored it.

chimney off his own lorry, which he then had to return to the depot, where he was disciplined for having lost his chimney. Truly a very different world. Very few lorries survive from the conflict; one, a Foden tipper, has been restored to its War Department khaki green and has appeared at several commemoration events in recent years where it attracts much attention and respect.

## Did you know?

Army Fodens were fitted with sprung buffers on the front, like railway engines, so they could come up behind a lorry in trouble on convoy and push it forward or out of the way.

Many of the established steam builders had a go at making steam lorries, but new names grew up as well – some looking surprisingly like existing designs. Companies like Sentinel, Yorkshire and Atkinson never made anything other than steam wagons and lorries, whereas Allchin, Wallis & Steevens, Tasker, Burrell, Clayton & Shuttleworth and Aveling & Porter all had a go at overtype steam lorries, making them in some number. By and large these were conventionally coal-fired with horizontal boilers, but there were many other makers, and their geographic spread is quite fascinating. The backgrounds of the companies were also varied,

**The First World War was a large user of steam lorries for delivering supplies to the Front. Many events were held to commemorate the centenary of the beginning of the conflict, and the surviving 1917 example attended 'Tracks to the Trenches' in Staffordshire in September 2014.**

and not necessarily in steam or agricultural engineering. Some were more successful than others, and it has been suggested by some historians that few or no examples of some makes survive because they were excellent products. Being so good meant that they were worked extensively, wore out more quickly and thus were replaced with internal combustion lorries when the time came, rather than another steam vehicle.

Clarkson of Chelmsford made lorries and buses which looked much more like internal combustion vehicles under the National Steam Car (1911) Ltd but for a chimney. Likewise, Liquid Fuel Engineering (Lifu) took advantage of not needing space to store solid fuel like wood or coal, and a much lighter and compact vehicle was the result, although not suitable for heavy loads in the way that something like a Foden or Sentinel was.

In the immediate post-war era, several manufacturers who had previously been producing overtypes switched their focus to undertypes, attempting to compete with Sentinel. Among these companies were Clayton and Garrett. In 1922 Foden began producing the celebrated C-type overtype. It was not a revolutionary wagon by any stretch of the imagination but had improvements such as a better driving position and the option of a windscreen. This is something we take for granted now, but at that time it was a real move forward in driver comfort. In 1923, Sentinel launched a much-updated wagon, the 'Super' Sentinel. In 1924, Fowler made their attempt to enter the undertype market. The various undertypes of the time were frequently fitted with windscreens to improve crew comfort. In the early 1920s, in an attempt to circumvent the weight regulations of the period and allow a higher capacity, several companies had experimented with the idea of an articulated trailer. With the brake and tyre technology of the era, such designs were often found to be difficult to control, with a propensity for jack-knifing. In 1926, Garrett produced a rigid six-wheeled wagon. Both Sentinel and Foden quickly brought out six-wheelers, and these became a large percentage of the output of these manufacturers for the remainder of steam wagon production. Around this period, Foden made several attempts to build undertypes, with the E-type being largely a failure. Yorkshire produced the updated WG, WH (shaft drive) and WJ (six-wheeler) wagons.

Fowler steam lorries are represented in preservation by a solitary example. As noted above, it is said that there are two reasons why few examples survive. These are that either the machine was very good and thus worn out after a hard-working life and not worth saving, or that the lorry wasn't a great success and thus not used much, meaning it lasted longer than

***Right***: The Great Dorset Steam Fair is one of the major heritage events each year. Many lorries visit, but not many have trailers as this Foden 6-ton example did in August 2017.
***Below***: Preservation has resulted in some amazing rebuilds and restorations. April 2018's Great North Steam Fair was graced by the Worbey family's Garrett overtype, one of the missing links of steam heritage which can now be enjoyed once more.

Mick Walters' Garrett undertype attends many events in Yorkshire under its own steam. In June 2012, it was appearing at the 200th anniversary of the Middleton Railway in Leeds.

the others. The jury is out with Fowler lorries, but the crew members of the survivor enjoy its ability to cruise at 20 miles an hour. The prop shaft drive is also an innovation in advance of Sentinel moving from chain drive to follow suit. Certainly, the Fowler turns heads when it appears at events or on the road. The Fowler only survives having been cut up at a scrapyard in Leeds with the parts remaining in the yard long enough for the legendary preservationist Tom Varley to discover it and take on the challenge. Built as a gulley emptier, it was the subject of a giveaway diecast and plastic model used as a promotion by Hall's Mentholyptus, whose livery the full-sized lorry carried for a time in the 1980s.

Tom Varley established a reputation for bringing steam lorries back from the brink of oblivion. His restoration of a Fowler steam lorry remains unique, and the sole survivor, rescued in the late 1960s, is still with his family. In September 2019, it was exhibited at Old Hall Farm, Bouth, Cumbria, which is now the base of the reconstituted John Fowler & Company Ltd.

## Did you know?

During the latter part of the 1920s and into the 1930s, several Sentinel and Foden steam lorries were given extended leases of life by being converted into railway locomotives. Somewhat unconventional in looks, the slow speed and high power of the geared engine were ideally suited to the shunting locomotive application on industrial railway systems in factories and works. Ironically, Sentinel then began to branch out into locomotive building proper, using lorry boilers and engine units initially to modernise old conventional steam locomotives and then construct new-build bespoke locomotives. The construction of steam locomotives by Sentinel continued well into the 1950s, such was the quality and ease of use of the products.

By the beginning of the 1930s, the landscape was becoming difficult for steam wagon production. Many manufacturers had gone out of business, and many others had turned to internal combustion. The only major manufacturers who produced new designs in this period were Foden and Sentinel. In 1930, Foden launched their revolutionary O-Type, better known as the Speed Six and Speed Twelve wagons. They were a valiant attempt at producing a modern steam wagon, but suffered from reliability problems, primarily due to issues with the boiler design. At the end of 1932, Foden turned to diesel wagons, but single examples of the Speed Six and Speed Twelve lorries survive in preservation and both were launched at the 2017 Great Dorset Steam Fair after extensive restoration and many years of hard work. The performance of these two lorries has been eagerly observed to see quite how they fare compared to the historic accounts of their operation.

Back in Shropshire at the Sentinel Waggon Works, research continued. The American steam power proponent and builder of successful high-quality steam cars Abner Doble joined forces with Sentinel in July 1931 to experiment and innovate into the future of commercial steam transport. At the same time, Doble's brother was working similarly with Henschel's in Germany, where an S6 lorry was sent on test. Back in the UK, a prototype Sentinel-Doble lorry was tested the following year with a two-cylinder compound double acting engine in situ on the back axle. At the same time, a new automatic stoker was fitted to certain DG-type waggons to make them one-man operated. Given the success of other fuel sources with steam cars,

Two epic restorations of the last thirty years are the ground-breaking Foden Speed Six and Speed Twelve lorries. Both have passed through several hands over that period, but the restoration of each was recently completed by Mike Dreelan of Aberdeen, and he can be rightly proud of his achievement. The lorries made their heritage debut at the 2016 Great Dorset Steam Fair, as shown in this photo.

it's perhaps surprising that coal held sway with Sentinel right to the end of their production. Doble continued to try out concepts throughout the early 1930s, but with Sentinel running into financial trouble in 1936, all future development was halted and Doble returned to America. It has been suggested that Doble and his experiments were one of the reasons that Sentinel found themselves financially precarious; certainly after Doble left, Stephen Alley bought back into the company and it returned to a stable state. Throughout the 1930s, Sentinel were active in both the steam and diesel lorry market, both enjoying very good reputations.

By 1933, Sentinel were the last major players in the United Kingdom remaining faithful to the production of steam lorries. In that year, they created the ultimate steam lorry that would be seen on the UK's roads for the next two decades. This was the well set-up late model Sentinel S-type with underfloor four-cylinder engine, two-speed gearbox and worm drive axle, which could easily hold its own in the traffic of the day. On pneumatic tyres with electric lights, the acceleration was good and the payload realistic, challenging both the diesel competitors and the legislation that had been introduced as a result of the Salter Report – which will be referred to later as it ushered in the end of the commercial use of steam on the road. However, it still required steam raising and a ready supply of coal and water. Introduced on to the market in 1933, the S-type was a fast and well thought out design that proved to be reliable. Its stylish looks and modernised cab with set back boiler even now lead to observers sometimes thinking that an S-type is a diesel lorry. It was available in four-, six- and eight-wheel form, with the designations S4, S6 and S8 accordingly. Despite the looks and performance, steam was on the wane commercially, and in 1938, production of steam Sentinels ended. Sentinel had made internal combustion vehicles from 1933 to 1936, having bought Garner Motors, but began making their own design in 1947; as with the steam lorries, they were very superior products. These lorries were quite radical at the time, featuring an underfloor horizontal diesel engine, all-steel monocoque cab featuring internal sliding doors, bench seat and concealed radiator.

A last hurrah for Sentinel was the construction of some 108 S-type steam lorries produced for the Argentine government in 1950 (of which two still survive in Argentina) and the last for the home market was built in 1952 as a prospective order for the National Coal Board. Whilst the latter performed well, the NCB did not take up the minimum order required by Sentinel and it was converted to a diesel unit in a very short time. Sentinel's production went completely over to diesel lorries. Construction of the Sentinel steam locomotives continued a little longer until 1958. The firm had then become part of Rolls-Royce, which meant that spares became very hard to obtain, a further factor in the demise of steam as a viable force on the roads.

A superb restoration of a Sentinel S4 is this example, fresh out of overhaul at the 2019 Shrewsbury Steam Rally. The layout and look of the S4 looks modern even today compared to the Foden overtypes of the same period.

# 3

# Steam at Work and on the Wane

The steam lorry enthusiast is fortunate to have some excellent first-hand accounts of working with the vehicles, either in the magazines of the Road Locomotive Society or the National Traction Engine Trust, but also in some books by the late Jack Hampshire and the late Ken Judkins. The latter had worked as a demonstrator/delivery driver for Foden and Sentinel and described his experiences vividly. Correspondents later would take issue with some of the tales Ken had told, but nonetheless, his books *My Life in Steam* and *More of My Life in Steam* are well worth seeking out. Jack Hampshire worked with many types of steam engine during his career and his practical tales of having operated different lorries are better than any theoretical account. Both authors were active in the preservation era as well, and their knowledge and experience were valued by many.

## Did you know?

Very few film records remain showing steam lorries in everyday service. One piece of popular film on YouTube is 'The Open Road' by Claude Friese-Greene, who was filming Britain in the 1920s in colour. This particular film dates from 1926 and is devoted to the streets of London, and the first sequence shows a typical scene with trams and buses vying for space on the roads. Amongst the traffic seen waiting for a right of way is a Sentinel steam lorry, laden high with sacks of who knows what? Although on screen for a split second, it's a wonderful snapshot of working steam.

Road steam disappeared through becoming uneconomical to operate, and unpopular with British governments. By 1921, steam tractors had demonstrated clear economic advantages over horse power for heavy hauling and short journeys. However, petrol lorries were starting to show better efficiency and could be purchased cheaply as war surplus; on a busy route a 3-ton petrol lorry could save about £100 per month compared to its steam equivalent, in spite of restrictive speed limits and relatively high fuel prices and maintenance costs. The picture was the same in agriculture: the basics of pounds, shillings and pence stacked up against the steam engine.

Throughout the 1920s and 1930s successive governments placed tighter restrictions on road steam haulage, including speed, smoke and vapour limits, and a 'wetted tax', where the tax due was proportional to the size of the wetted area of the boiler; this made steam engines less competitive against domestically produced internal combustion engined units (although imports were subject to taxes of up to 33 per cent).

Steam lorries found widespread use until the late 1930s. A Mann example was at work on road-making duties in this image, repeated across the country with a variety of makes.

The versatility of the steam lorry before the Salter Report is shown in this London-based Foden 3-ton example with trailer. Once commonplace, they were pretty much wiped out within a decade.

Arthur Salter, a journalist and author, chaired an influential conference of road and rail experts in 1932 which reported in 1933. The report directed British government policy for transport funding for decades to follow. The widespread use of public road passenger vehicles and private motor cars was seen to be of great benefit, but it noted that the growing numbers of motor vehicles were capable of making many more journeys than the previous generation of horse-drawn traffic. In particular, the hauling of heavy loads by road was useful but expensive as it caused more wear on the carriageways. To counteract this, the report recommended that local authorities should be able to restrict heavy traffic from local roads and should be relieved of the burden of funding their maintenance. Instead, the motor vehicle should fully pay its way. The report recommended replacing the established system of road funding from the local ratepayer wholly onto the operators of motor vehicles through changes in road pricing.

As a result of the Salter Report on road funding, an 'axle weight tax' was introduced in 1933 in order to charge commercial motor vehicles more for the costs of maintaining the road system, and to do away with the perception that the free use of roads was subsidising the competitors of rail freight. The tax was payable by all road hauliers in proportion to the axle load; it was particularly damaging to steam propulsion, which was heavier than its petrol equivalent. Steam lorry use reduced massively, and many machines were laid aside after a short working life. Others were converted to pneumatic tyres and went on to be operated commercially for another decade or so. Solid tyres remained for some lorries used inside works, such as the Brown Bayley Steelworks Sentinels in Sheffield that lasted until the 1960s. Some tar sprayers remained on solids as well, as road-making plant was exempt from duty, though latterly the last working examples of these also were on pneumatic tyres. The steam was also hot enough to melt coal tar for spraying. The North Thames Gas Board was operating coke-fired steam lorries after the Second World War, whilst several of the Sentinel S4s were put back into operation during the Suez crisis of 1956.

Initially, imported oil was taxed much more than British-produced coal, but in 1934 Oliver Stanley, the Minister for Transport, reduced taxes on fuel oils while raising the Road Fund charge on road locomotives to £100 a year, provoking protests by engine manufacturers, hauliers, showmen and the coal industry. This was at a time of high unemployment in the mining industry, when the steam haulage business represented a market of 950,000 tons of coal annually. The tax was devastating to the businesses of heavy hauliers and showmen and precipitated the scrapping of many engines.

Atkinson undertype lorries from the fleet of Bibby's in Liverpool laid up at the end of their working lives.

The new charges were blamed for driving heavier steam traction off the road in favour of the lighter lorries powered by internal combustion engines using imported oil. It also introduced an unpopular requirement for hauliers to produce paperwork to show that their drivers were operating safely and not being overworked. Previous attempts to impose safety rules on hauliers had been nearly impossible to enforce. The threat of withdrawing a licence was viewed as the best way of ensuring the new regulations were respected by the industry.

It has been seen how technical innovations were employed throughout the development of the steam lorry, particularly as the law and technology changed hand in hand throughout the 1930s. The basic layout of the steam lorry of that period may have been like that of the 1900s, but the details were so very different, with power brakes and automatic stoking as well as the mechanism of power transmission itself and enclosed engines. There was one thing that the steam builders could not do – the work for the driver and crew remained hot and dirty, with preparation time that the diesel and petrol lorry did not need, and the human factor was the ultimate nail in the coffin for the steam lorry.

The view from the cab of Richard Straughan's Sentinel DG8 as it runs down the main street in Shildon, County Durham, in May 2018. The boiler is to the fore in the picture and the handle for the firing chute is clearly visible.

The crew of the steam lorry were not the only people adversely affected by the late 1920s. In 1929, the Secretary of State for the Home Department was asked:

> Whether his attention has been drawn to numerous accidents which have occurred recently in London streets owing to collisions occurring between vehicles through drivers being momentarily blinded by steam and black smoke being emitted from passing lorries or steam wagons; can he state whether any police regulations in the Metropolitan police district exist to deal with this danger; and in how many cases have prosecutions taken place against owners of such vehicles who have been responsible for this nuisance during the twelve months ended to the last convenient date?

In reply, the Secretary of State, Sir William Joynson-Hicks, noted that due to the difference in regulations in different parts of London, it was hard to quantify the actual number of accidents, but the previous twelve months had seen 134 proceedings taken to case. There were also accidents to pedestrians recorded where they had not been seen due to steam and smoke obscuring the drivers' vision.

## Did you know?

The taxation challenges saw some steam lorries have ridiculously short working lives – some of which were less than five years. Barely run in machines were sent to the scrapheap as their value plummeted. It is said that some were used in their owners' locations as landfill, such as the Bardon Hill quarries in Leicestershire. An incredible occurrence was the laying aside of the fleet of W. J. King of Bishops Lydeard in the early 1930s, but the lorries weren't scrapped, instead surviving derelict until 1988 when an amazing auction was conducted, and the steam fleet sold into preservation for restoration.

Steam power on the main line railways of Britain finished in 1968; industrial railway steam continued into the 1980s. Whilst the majority of road and agricultural steam had ended by the early 1960s, as with the railways, there were pockets of survival where steam was still used out of loyalty, sentimentality or downright practicality. A throwback to the earliest Sentinel Standard waggons was their continued use at Brown Bayley's steelworks in Sheffield for moving steel castings internally, where their rugged construction and solid tyres coped with the

**The end for so many steam lorries and tractors: a derelict Foden awaits its inevitable fate.**

rough conditions in the works. Whilst many were no longer registered for road use, the waggons were at work into the late 1960s and survived to be preserved by museums and enthusiasts. Photographs of the lorries in use in the 1960s show just how much neglect and abuse the steam engine could take compared to its internal combustion cousins. Their commercial use was only bettered by the Lloyd-Jones brothers in North Wales and Dave Coppard's former Glossop's tar sprayers. The Lloyd-Joneses worked consistently until the early 1980s, when the waggon was laid up and eventually entered preservation, now being seen on the roads and rally fields of Shropshire and North Wales, not too far from its commercial operating area.

***Left***: Brown Bayley Steels in Sheffield were late users of steam lorries, Mike Squire recording this Sentinel Standard in the works on 17 May 1969.
***Below***: The late Mike Jones captured Lloyd-Jones Brothers' Sentinel tar sprayer and Fowler steam roller at work at Llanbedr Hall, Llanfair DC, in 1973.

## Did you know?

Ewan and Murray Lloyd-Jones of Llanfair DC in North Wales ran their steam tar sprayer and Fowler steam rollers into the 1970s and 1980s alongside more modern plant, such as JCB excavators, but they still did an excellent job of work. Colour photographs of the operations are vivid records of working steam within living memory for many. By this time the bitumen that was used on road works caused problems as the steam from the lorry was not hot enough to melt the substance for use.

Steam lorries were relatively easy to get rid of: the bodies could be burnt, and the rest disposed of as scrap metal – or recycled in another way. Some Foden engine units saw further use as portable steam power plants, and a more widespread use of boilers was as soil sterilisers or heating for greenhouses or pig swill. Enough have survived into the present day for some of these boilers or engine units to be rebuilt into complete lorries again, testament to the skills of the restorers who saw the challenge and took it on. As with life-expired vehicles since the start of mechanised transport, many were laid aside in fields and yards, or with dealers who could not sell them. These were not scrapped immediately, being rescued ultimately by enthusiasts but not before they had sunk into dereliction, making the task of restoration even more significant.

At the premises of Pip Atherton near Widnes, the body of Foden UR 1328 meets a fiery end.

The boiler and engine unit of UR 1328 survived in other use, long enough to be preserved and rebuilt by Vince Allen, making its debut at Beamish Museum's Power of the Past event in May 2007.

Even in preservation, vehicles can fall in and out of use. This is a Sentinel DG4 tractor, rebuilt by Sentinel from a DG6 lorry in the 1930s and stored on a farm in Kent in 2008. By 2019 it had been overhauled and was in steam with an owner in eastern England.

## Did you know?

One Foden boiler and engine unit was removed from a lorry and spent some time as an anchor in a harbour!

The most incredible restorations have occurred in recent years. In April 2012, a Foden lorry was shown at the Great North Steam Fair which had left the site over a decade before as a boiler and a damaged set of cylinders. There had been suggestion that the boiler had at one point been used as an anchor on the Isle of Man!

# 4

# Preservation of a Full-Size Lorry?

When the first steam traction engine race became a rally at Appleford in the early 1950s, it wasn't long before a steam lorry joined the ranks of vehicles taking part. Many lorries were either still at work or only recently retired and in scrapyards, and it seems odd to think that some of the Sentinel and Foden wagons were just twenty years old. Edgar Shone of the Crown Hotel, Cricklewood, London, became a specialist collector of Sentinel waggons and several lorries in preservation still bear the name 'Crown Steam Haulage' on the front as testimony to his ownership. Over a hundred Sentinel steam waggons survive in preservation in 2019.

Typifying how steam lorries entered preservation, this Sentinel S-type and DG were exported to a Canadian enthusiast in the late 1950s and returned to the UK in 2014. They are now under restoration at Statfold Barn Narrow Gauge Trust in Staffordshire and their return to the steam rally scene is eagerly awaited.

In May 2016, a major sale was conducted of the collection of the late John Keeley at Knowle Hill, Berkshire. A pair of Sentinel S types are being inspected on the viewing day before the auction. Both were sold to enthusiasts who are thoroughly overhauling the lorries after their many years in store.

Steam on the road is always a splendid spectacle. The Sentinel DG6 of Edgar Shone was taking part in the Historic Commercial Vehicle Society run of 1971.

## Did you know?

The first recorded steam lorry to attend the famous Appleford steam rally was a Sentinel S-type tipper, owned by Dick Woolley who had driven it over from Shropshire for the day in 1952. The lorry was still being used for work at that time!

As the preservation of steam lorries moved on through the 1960s, the search began for rarer and more unusual machines. *Dorothy*, a Thornycroft machine, was one of the first to appear on a rally field in the ownership of John Crawley, its 1906 outline being a massive contrast to the Fodens and Sentinels widely seen at that time. The late Tom Varley of Gisburn had both Foden and Sentinel steam lorries, but he sought out scarce steam lorry remains in Leeds scrapyards, recovering examples of both Yorkshire and Fowler wagons and rebuilding them patiently and displaying then at his caravan park. Many others would have given up, as the chassis had been cut up into pieces, but Varley persisted with seeking parts and reconstructing patiently. He also acquired a very early Yorkshire lorry of 1902 and rebuilt it, a superb link back to the first years of steam commercial vehicles with wooden wheels and iron tyres. Having established a name for himself for restoring interesting steam vehicles, he then moved to scour Australia for unusual survivors. Amongst a number of other road

The late John Crawley was instrumental in saving and preserving the rare Thorncroft lorry *Dorothy*, which became a very welcome addition to the 1960s steam rally scene, as photographed by the late Tom Charman.

engines, he acquired Foster, Atkinson and Aveling & Porter machines from that country to restore them and show them on the UK rally scene. Once restored, these lorries are now unique in the UK, and there are precious few survivors in the rest of the world – mostly still in Australia. Whilst the collection was dispersed after Mr Varley's passing, the Fowler remains in the family ownership and can be seen in steam in Cumbria on occasion. The other lorries in the collection are spread throughout the UK and enhance the rally scene wherever they are displayed.

The preservation story has seen engines move widely across the UK and the world. Tom Varley blazed the trail with repatriating vintage steam, but he was not alone. A Ransomes, Sims & Jefferies steam lorry power unit was returned from Australia to the UK in the 1990s and formed the basis of a rebuild into a wagon by enthusiast Pat Jenkins. Completed in the early 2000s, it was a real head turner on the rally scene, as no-one had seen a Ransomes lorry in England since the 1940s at least. Having restored the Ransomes, Pat then turned his attention to a rare Mann – not unique, but one with a fantastic story as it had been operated by the Manchester locomotive builder Beyer Peacock, who themselves had put a toe in the steam lorry market before the First World War.

There has also been a flow of lorries in the other direction out of the UK, in addition to commercial exports in working days. Preserved Fodens have gone to Germany, Sweden and the USA to name but three; similarly with some Sentinels – and a tar sprayer has gone as far as Australia, to join the ranks of steam lorries already there. An invaluable source of information is the already-mentioned *Traction Engine Register*, which despite its name

***Above***: Even in the 1960s and 1970s, treasures remained to be found, such as this chassis of a Yorkshire lorry found in North Wales by Mike Jones of Ruthin.
***Below***: Tom Varley took on the challenge of rebuilding the 1905 Yorkshire from a collection of remains including the chassis. It is now owned in South Yorkshire and appears regularly at events across the north of England, including the Great North Steam Fair as here in April 2014.

***Above***: An unusual venue for an unusual steam lorry. The 2017 charity event CarFest North in Cheshire had steam engines for the first time, one of which was Mark Hitchin's Mann lorry, which once worked for the locomotive builder Beyer Peacock of Gorton, Manchester.
***Below***: Two Mann steam lorries at the Bedford Steam Rally in 2005 when it was held at Old Warden, home of the Shuttleworth Collection.

includes steam lorries both in the UK and vehicles exported in the preservation period. Imports are also listed under their original manufacturers.

As the years have passed, the ranks of preserved lorries have been joined by Mann, Ransomes and others. At the present time, a Burrell steam lorry is under reconstruction in Sussex – a very significant gap in the ranks of surviving lorries. The 2019 Great Dorset Steam Fair saw the first public appearance of a Clayton & Shuttleworth steam lorry that has been stored away from any access for many decades. It has since moved on for restoration and will no doubt grace the rally fields and roads of Britain in future years. The preservation scene cannot be regarded as dull – and the colourful variety of liveries worn by Sentinels over the years is borne out by many of the restored examples that have been shown over the last decade or so. At the other end of the spectrum is the incredible survivor that was in the Henry Ford Museum in America for many years. Henry Ford himself took a fancy to a 1913

The engine unit of the Burrell steam lorry currently under rebuild and repair was shown in the Road Locomotive Society marquee at the 2018 Great Dorset Steam Fair.

Summer sunlight shows the original paintwork of this Foden lorry to advantage when attending the Beaulieu Steam Revival in June 2010. It was displayed from the 1920s to the early 1980s at the Henry Ford Museum, Dearborn, USA, and is in remarkable condition.

Foden lorry that he saw in Manchester when touring the UK in the 1926 and bought it for his museum in Dearborn. It remained there until 1980 when a collections rationalisation deemed it surplus to requirements and it returned to the UK. Since that time, it has undergone a very sensitive overhaul and recommissioning, retaining its original paint, which has been carefully conserved, giving us a rare insight into what a working steam lorry from the 1920s would have looked like.

## Did you know?

Carrying the name *Britannia,* a 1926 Foden C-Type 6-ton steam lorry is famous for its 'Grand Global Tour' of 1968 to 1972, when it became the first steam-powered road vehicle to circumnavigate the globe. Owned since the 1960s by enthusiast Michael List-Brain, it is the only steam vehicle to have achieved this. In its time *Britannia* has travelled in places as diverse as Afghanistan, Australia, India and the United States, and it has seen the likes of everybody from Presidents to pop stars travelling in the cab.

What's it like to drive a steam lorry? Depending on whether it's an overtype or undertype, Sentinel, Foden, Mann or other make, it's still a steam engine and the basic principles of controlling the steam boiler remain. However, a Foden lorry from the 1900s is very different to a Sentinel S-type waggon. The author cannot claim to be anything of an authority on driving a steam lorry – he has driven the lightweight Sentinel waggon No. 1465 a few miles and a Mann lorry a couple of times at rallies and shows, but other than that, has simply observed. The Sentinel was a great experience with steam brakes for someone used to an old steam roller or traction engine. It's performance on the local bypass when it began to superheat was exhilarating to say the least and a real eye-opener to how the steam lorry manufacturers really were trying hard to equal, if not beat, their petrol and diesel competitors.

The engine and driving position of a Foden, a great place to hide from the rain during an October 2019 road run in Worcestershire!

***Above***: Driver's view from Terry Varley's Fowler undertype at Old Hall Farm, Cumbria, in September 2019. ***Left***: A meeting of road and rail at Locomotion, Shildon, County Durham, in May 2011. Lightweight Sentinel No. 1465 stands close to the famous *Hogwarts Castle*, which was giving rides that weekend.

Firing a Sentinel is another game the author has played on occasion, and it's very odd for someone used to shovelling coal onto a fire grate one can see to be tipping coal down a firing chute to an invisible fire bed and firing by intuition. Lorries with compound engines have the facility often to put high-pressure steam into both cylinders to aid starting or on hills, and that's a skill to master, driving and switching between these on something like a Foden.

Lorries under restoration aren't often seen in public. Tom Charman caught this S4 part restored at Wistaston Rally in 1964. The water tank and boiler behind the crew can be clearly seen.

In 1994 the Great Dorset Steam Fair held a special themed display of steam lorries. In other years, it has had shows of makes or other types of engine such as portable engines or steam rollers, but for one year it was the turn of the steam lorry, from Atkinson to Yorkshire and with many unusual makes and models in between. Static and steaming, the gathering was unprecedented and has never been repeated, with engines appearing from all over the UK, including Northern Ireland, and several being brought out from museums specially. These included the Tasker and Wallis & Steevens lorries from Hampshire Museums and the Leyland from the British Commercial Vehicle Museum in Lancashire. It's arguable that the time has come for another gathering of this type as in the two decades or more since the event, so many more unique and interesting lorries have entered the heritage scene.

White is not necessarily the most practical livery for a steam lorry, however authentic it might be! Carrying Hovis Bread's colours, this Sentinel charges up the hill at the 2013 Dorset event.

## Did you know?

The Engineering School of HMS *Sultan* in Gosport, Hampshire, has its own Super Sentinel lorry, maintained by students and officers. It's often to be seen at steam events across the south of England.

With a sea of caravans as a backdrop, the Super Sentinel of HMS *Sultan* is a regular attender at the Great Dorset Steam Fair and was arriving on site in 2013 in advance of a full week of activity.

# 5

# What Next? The Steam Lorry Today

Traditionally, if one wanted to see an example of something kept for posterity, a museum would usually be the first port of call. The UK is no exception, and the Science Museum collection includes a Sentinel waggon and a Foden lorry based at the National Collections Centre near Swindon in Wiltshire, whilst their Leyland lorry is on loan to the British Commercial Vehicle Museum. The latter museum also has a Foden steam lorry on display as well as the earliest steam van by Thornycroft. Several municipal or local authority museums have steam vehicles with local associations, like Glasgow Riverside with a Sentinel waggon and the locally built Thornycroft, Tasker and Wallis & Steevens lorries in Milestones in Hampshire. Elsewhere, ThinkTank in Birmingham has an old Foden overtype, Liverpool has a Sentinel tractor and there are others, mostly with local connections in that they were used locally or presented by local enthusiasts.

Currently not on public display, the 1916 Foden lorry *Pride of Edwin* is part of the Science Museum collection and will be accessible when the National Collections Centre at Wroughton opens in 2023.

The British Historic Commercial Vehicle Museum at Leyland fittingly displays this Leyland undertype steam lorry. Also part of the Science Museum collection, it has been on loan to Leyland for many years.

Glasgow's Riverside Transport Museum would not be complete without a Sentinel steam waggon. The example on display was built in 1916 and was one of the late survivors of the Standard type that worked at Brown Bayley Steels.

Liverpool has a Sentinel DG4 tractor on show at the Museum of Liverpool, Criddle & Co. being a long-established firm who operate in the city and ran a fleet of Sentinels.

**Local manufacturer Taskers of Andover are celebrated in the Milestones Museum of Hampshire History in Basingstoke. The now-unique Tasker steam lorry is to be viewed there, alongside a more conventional product of the firm in the Economic traction engine.**

The North of England Open Air Museum at Beamish in County Durham runs the Great North Steam Fair each spring. This four-day festival encompasses the whole museum site and usually features several steam lorries, with an eye to the interesting, old and unusual. Within the museum are a recreated street and several other settings, like a colliery village and pit yard, so scenes from the past are re-enacted. The event is perhaps the closest the present-day enthusiast will ever get to seeing steam lorries in the urban settings they would have been found in with associated traffic and street furniture. It's well worth trying to visit when the event is on.

There are also steam museums and collections with lorries in their displays, notable ones being Strumpshaw and Thursford in Norfolk. The famous Bressingham Steam Museum parted with their lorries a few years ago sadly, but there are others around the country. Many are listed in the *Steam Heritage Yearbook* and in the *Traction Engine Register* – the latter publication lists all known surviving road steam vehicles, make by make. It's a very useful book to have, and shows just how widespread the survivors are, as well as identifying the odd and unusual. Barrie C. Wood also produces the *Traction Engine Museum Guide* on a regular basis and for less than ten pounds, it's well worth the investment.

Some of the museums will have their own archive and photo collections, which are great for research. On the internet, many County Record Offices have online catalogues of images and records which can be researched, whilst other locations hold specific archive collections relating to operators and manufacturers. The Museum of English Rural Life at Reading has

***Left***: Beamish Museum in County Durham provides the fabulous setting for the Great North Steam Fair. April 2017's event saw the appearance of Alfie Cheyne's fine S4 Sentinel lorry at the Pockerley waggonway end of the site, with a reconstructed church in the background. ***Below***: The town street at Beamish is a great location to view three Sentinel steam lorries on an April 2017 evening, headed by the Cook family's 1931 DG4.

notable holdings of Fowler, Ransomes and Wallis & Steevens material amongst others and one can spend hours engrossed in original documents, photographs and register books. Other online archives are available to explore, from private individuals, publishing houses and interest groups. With the never-ending rise of social media, there is also a steam wagon Facebook page, adding to the material accessible to the steam lorry enthusiast.

Away from the actual vehicles themselves and parts thereof such as badges and worksplates, some of the most tangible remains are the factories and workshops where the lorries were built. Several have been afforded protection through listing and a few have even been given blue plaques by civic societies or red wheels by the Transport Trust. In Glasgow, the design offices and pattern shop of Alley & Maclellan Sentinel Works is a unique survivor. It was built on the Hennebique principle in 1903–04 and was the first fully reinforced concrete building, having a ferro-concrete frame and panels, and the third oldest to survive in the UK. This building's form anticipates the American-inspired daylight factories by Albert Kahn, unlike any similar building in England. Sadly, the building is derelict despite being Grade A listed in the Scottish listing system and it is to be hoped that a positive future is in sight for the works. It is not just of architectural interest but is noteworthy for what was made there and the developments that took place in the steam waggon business thereafter.

The Sentinel Waggon Works moved south to Shrewsbury in 1915 and although the company is no more, the works building remains with the company name proud above the gateway along with the large Sentinel figure. It has become a place of pilgrimage for steam enthusiasts, and many restored and rebuilt waggons have revisited the works to have their photographs taken by proud owners. Shrewsbury even has a pub named the Steam Wagon Inn to commemorate the heritage of the town.

**The frontage of the Sentinel Waggon Works in Shrewsbury survives to this day as a fine piece of industrial heritage and celebrates the engineering history of the town.**

In Leeds, the buildings of both the Mann Patent Steam Cart & Wagon Company and the Yorkshire Patent Steam Wagon Company survive to this day. Mann's is still in use as an engineering facility for Taylor's, another long-established Leeds company, and is the most complete remaining example of one of the former Leeds engine works. In recent years the Yorkshire Steam Wagon works has been taken over by Pickersgill Kaye, a specialist lock maker. Amongst the products made by the firm are many door components for the railway industry. One of the biggest contracts won by the firm was the order for emergency door release handles for Hitachi's Intercity Express trains for the Great Western and East Coast Main Line! The Leeds Civic Trust in connection with the Leeds and District Traction Engine Club have in recent years been active in marking the steam heritage of the buildings with blue plaques and ceremonies at which examples of both Mann and Yorkshire lorries have been present.

Models provide an affordable way to enjoy steam lorries in your own home, or with others in the case of working steam. On the smallest scale, there are metal and plastic kits of Fodens and Sentinels for self-assembly. These have been made in the past by Anbrico, Keil Kraft and others, whilst Langley Miniature Models and Wills Finecast cater for the current enthusiast. The plastic kit ones fit with the popular model railway scale known as 00 and can be picked up even now for less than £10. The Matchbox Models of Yesteryear series originally had a Sentinel steam lorry in the initial 1960s series and from the 1980s featured Fodens, Atkinsons and Garrett lorries. In more recent times, the Corgi Golden Age of Steam diecast model series has included larger and quite detailed Fodens and Sentinels, a number representing well-known historic liveries or lorries that can be seen on the rally scene across the country. Smaller than these are the Sentinels and Fodens made by Lledo in their Days Gone range. Initially produced in historic colour schemes, there has latterly been a trend to make promotional models in colours of popular companies or products, more as a sales aid than a faithful recreation of history.

If you want to go larger in scale and investigate steam power, there are steam lorries in the live steam toy products of both Mamod and Wilesco. The Mamod steam lorry has been in production for over forty years, made in Birmingham, and is a great representation of a classic English overtype steam lorry. Fuelled by either meths or solid fuel tablets, they have been made in a variety of colours and make excellent display pieces when not in steam and being played with. Wilesco are a German make but their self-moving steam toys are very English in outline and design and are again fired by solid tablets. They make a Foden-style steam tractor, and also a vertical boiler lorry with the engine behind the cab; these are more expensive than Mamods in general. Slightly larger still are a range of gas-fired radio control Sentinel lorries made by Maxitrak in Kent. It's then a logical step to move on to a steam lorry that the owner can either ride behind or ride on. In the field of model engineering, there are many designs that one can choose to build or have built for you. In the scale of 2 inches to the foot, the undertype Clayton & Shuttleworth lorry designed by Robin Dyer is perhaps the smallest viable model that can pull an adult around. Since the design was published, many sets of drawings and castings have been sold and a good number of vehicles completed by steam enthusiasts around the world. Even today, finished lorries, unfinished models or sets of parts are often found for sale at auctions, on the internet or with specialist model steam dealers, such as The Steam Workshop in West Yorkshire or Station Road Steam in Lincolnshire to name two of the most reputable. Going larger, there are many Fodens, Sentinels and Atkinsons in 3-inch, 4-inch and 6-inch scale, the latter being very impressive vehicles that one can sit in rather than on or behind. The larger steam lorry models can be

A selection of Days Gone model steam lorries in various liveries – hundreds of versions have been made, and there's not space to display them all!

**Masham Rally in July 2015 saw a variety of 6-inch scale Foden miniature lorries, including the one seen here, owned by Jeff and Chris Hall from County Durham.**

registered for the road and are able to hold their own on the public road in terms of the speed they can manage just as well as their larger counterparts. There are even a couple of 9-inch scale models, which when seen on the road are almost as large as some of the smaller full-size Fodens. The ultimate is the reconstructions of long-lost vehicles in full size, and at the time of writing, the heritage world eagerly awaits the completed rebuild of a Burrell overtype lorry, the like of which has not been seen since before the Second World War.

Should steam not be a viable option, recently model ride-on wagons have been made which use electric mobility scooters to provide power but have steam generators to give the impression of what they are meant to represent at a fraction of the cost and much cleaner for the operator!

## Did you know?

The Disney film *One of Our Dinosaurs is Missing* was intended to feature a steam lorry used by the villains to steal the dinosaur skeleton from the Natural History Museum. The practicalities of using a real steam vehicle to film with overcame the production crew, so in the end a steam lorry was mocked up from a normal internal combustion lorry. To the layman, it's quite convincing, but to those who know, it's clear that it's not a real one as the boiler is tiny!

## Did you know?

The 1928 film *The Wrecker* features a spectacular crash between a passenger train and a Foden steam lorry stuck on a level crossing. The scene was filmed at Herriard on the Basingstoke & Alton Light Railway, in one take, and destroyed both the steam wagon and the South Eastern & Chatham Railway F1 Class locomotive.

Most steam rallies across the UK each year have at least one steam lorry, usually a Foden or a Sentinel. The County of Salop Steam Engine Society every year fields a particularly fine display of Sentinel steam wagons, many of which, as noted earlier, make a diversion and pilgrimage to the former Sentinel works to be photographed in front of the place they were made. Close by are many houses built at the same time for workers; the Sentinel village is but one of many planned industrial housing developments from the first part of the twentieth century laid out along the lines of a garden suburb. There have been several special occasions marked by the Shrewsbury event, and for certain anniversaries of either the company or the steam society, road runs have taken place, notably one taking in the former test route over the Horseshoe Pass near Llangollen. It's a real spectacle to see an

**Shrewsbury Rally is an annual event renowned for its gathering of Sentinels. This view along the engine line in 2019 shows nine lorries, not all of those attending!**

***Above***: The Sentinel Drivers Club have a presence at the Shrewsbury Rally, fronted in 2019 by a 1921 Standard lorry. Membership, sales and displays are all on offer to visitors to the stand.
***Left***: Road runs, whether organised or ad hoc, are a pleasant way to enjoy steam on the move. Sometimes, owners, crews and friends enjoy a meal at a pub on the way – as here at the Bay Horse, Middridge, in May 2018.

S-type Sentinel digging in for the climb and then roaring away from the summit – a true demonstration of the worth of these ground-breaking machines, even in the twenty-first century. An organised road run is also a splendid spectacle, allowing spectators to see a steam lorry at full speed on modern roads, a real sight to behold and quite often a surprise to modern road users who expect a vintage vehicle to hold them up rather than keep pace with them! At points along such runs or the end point, the vehicles are often parked up at a visitor attraction or a pub, and it's a great chance to meet and chat with owners and crews as well as seeing the lorries on the road.

As with many fields of specialist interest, there is much printed material to support further reading and understanding of steam lorries. The two monthly magazines *Old Glory* and *Vintage Spirit* cover steam lorries and the rally scene as they do the rest of the road steam world. Until the 1980s, the journal *Light Steam Power* also often had features on steam lorries and the works of Abner Doble, despite its concentration on steam cars and steamboats in the main.

Books wise, there have been dozens published, from reprints of original manufacturers' catalogues to many photo albums of preserved lorries on rally fields. In between have been collections of working day photographs, manuals and works photographs. Perhaps the most

Christmas gatherings are further examples, where steam is part of the fun of the festive season. Boxing Day 2007 saw Sentinel Super 7651 in attendance at the Station Inn, Kegworth; the event has now moved to Sutton Bonington with the closure of the inn.

useful and all-encompassing works have been Ronald H. Clark's *Development of the English Steam Waggon* and Maurice Kelly's two books *The Undertype Steam Road Waggon* and *The Overtype Steam Road Waggon*. Between the three publications, the full technical history of the steam lorry or waggon is documented, and the books will enhance any enthusiast's bookshelves despite having been out of print for over forty years. There is a twin-volume set on *The Sentinel* which also sets out the comprehensive history of the company and development of its various products, of which waggons were but a part. In terms of archive photographs, the John Crawley album *Steam Wagons in Focus* is an excellent book to seek out, as is *Kaleidoscope of Steam Wagons* by R. A. Whitehead, perhaps better known for his books on Garretts of Leiston and their products. Another from his pen, *Glossops – Kings of Tarspraying* is an in-depth look at the operators of a fleet of lorries fitted out for road making and an example of an excellent company history. Elsewhere in the world of publishing, there have been several histories of manufacturers and companies who made lorries, and sections of these books have been devoted to the vehicles – such as Michael R. Lane's books on Aveling & Porter, Burrells, Fowlers and Fosters.

Original catalogues, manuals and contemporary photographs are highly collectible and appear for sale with book sellers and auction houses from time to time. Failing that, the Road Locomotive Society and other publishers have produced reprints of some catalogues in recent years. Away from specific books on the vehicles and their maintenance, there are also excellent sources of information in some of the technical books from the period when steam lorries were commonplace. The 1904 publication *Steam Boilers – Their History and Development* has an entire chapter on what it terms 'motor car boilers' but this includes steam lorries, and is fascinating as it covers the boilers from that early developmental period when the Liverpool Trials were being carried out. Illustrations of the Lifu and Thornycroft boilers are particularly enlightening. Likewise, part works such as *The Book of Modern Engines and Power Generators* by Rankine Kennedy ran to six volumes, though only one covered steam lorries. For the modern enthusiast of historic books and source material, it is the volume that covers road steam which is the hardest to find! Moving on to other collectibles, from time to time, such things as engine makers' plates, name plates, lamps and other fittings are sold, privately or at auction, and make great display pieces.

A selection of reprints of makers' catalogues and handbooks.

***Above***: Companies sometimes used steam lorries in their own advertising, such as here. This image was used as a poster and as an enamel sign; both are highly collectable. ***Right***: Much as with twenty-first-century vehicles, drivers' manuals were produced to encourage the best practice in driving and maintaining expensive steam lorries.

Should the collection of artefacts not be in the reach or desire of the reader, but the wish to know more be sparked, research is not restricted to books old and new. Many of the manufacturers' records are kept in archives all over the country, mainly in locations where the makers were based – such as the Foden company archive residing with Cheshire Archives. The Museum or English Rural Life at Reading University also holds many records, including Wallis & Steevens and Fowler.

The National Traction Engine Trust and Road Locomotive Society, although general in their coverage of the preserved road steam scene, also often feature steam lorries in articles and photos. Of more specific interest are the Sentinel Drivers Club and the Foden Society, which cater for enthusiasts of those particular marques. Both have archives, sales sections and newsletters and it isn't necessary to own a lorry to be a member. Other products and subjects are covered including diesel lorries, but they complement the steam content and put the general company into context. The Foden Society have published a small number of specific books, including *The Foden Centenary Book*, which was written and intended for publication in 1956 but was not finally published and made available for another fifty years.

## Did you know?

In the more recent *Thomas the Tank Engine* story books, a steam lorry called *Elizabeth* is one of the characters, continuing the place of the steam lorry in children's stories into the twenty-first century.

The steam lorry was a commonplace feature on Britain's roads for five decades, and it's perhaps a sobering thought that the Sentinels that were only twenty years old when first preserved in the 1950s have now been in heritage longer than they were in commercial use. Despite their relatively short life, the survival of a couple of hundred steam lorries means that the sight of a steam engine keeping pace with modern traffic can still be seen each summer as lorries are taken to events and gatherings across the country. With great concern in the world now about climate change and vehicle emissions, will the steam lorry be seen again in commercial service in the UK? Whilst a traditional coal-fired vehicle is unlikely, who knows? Only time will tell.

The author's favourite steam lorry picture. Taken by Ian Cooper in May 2004, Sentinel tar sprayer 8122 leaves the rally field from the Tom Rolt Rally in Tywyn, which the author used to organise. It could date from any time in the previous six decades.